SCIENCE WITH A BEAT
BOOK THREE

WEATHER

Written by Jacquie Hawkins

WEATHER

CHAPTER ONE
HOW DOES A THERMOMETER WORK?

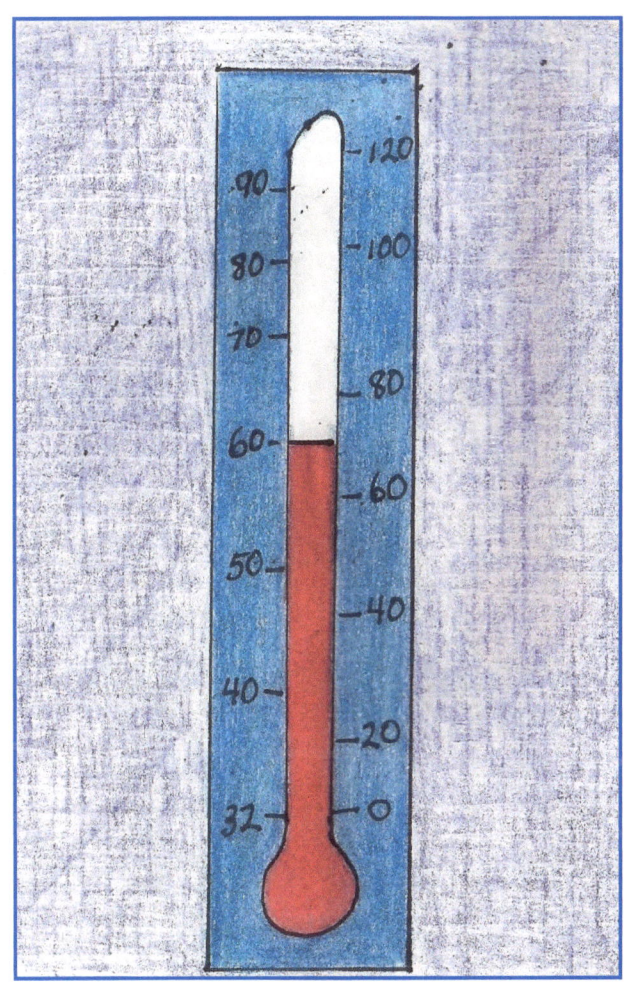

The <u>colored</u> or silverish <u>line</u> that you <u>see</u> <u>In</u><u>side</u> a ther<u>mometer</u>…

<u>Tells</u> if the <u>weather</u> is <u>hot</u> or is <u>cold</u>, And <u>it</u> doesn't <u>matter</u> which <u>one</u> you pre<u>fer</u>.

If the <u>day's</u> really <u>hot</u>, the <u>line</u> is quite <u>tall</u>. If it's <u>cold</u>, the <u>line</u> will be <u>short</u>. The <u>liquid</u> in<u>side</u> doesn't <u>care</u> which it <u>is</u>. <u>Its</u> job is <u>only</u> to <u>give</u> the <u>report</u>.

2.

Most of the time the liquid is mercury
But sometimes it is alcohol.
Both liquids expand when the weather is hot
And contracts when the cold weather calls.

Because of the tube that the liquid is stuck in
The liquid has no place to go
But up in the tube and down once again.
The weather determines the flow.

CHAPTER TWO
WHY IS THE GRASS WET IN THE MORNING?

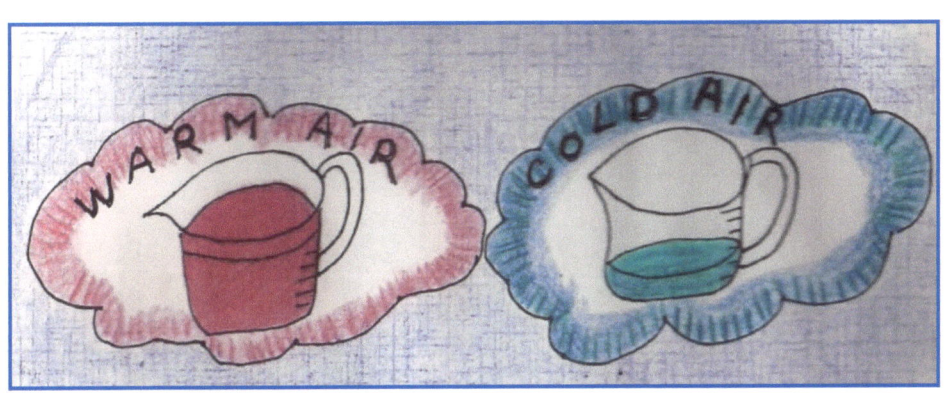

All through the <u>day</u> the <u>grass</u> soaks up <u>heat</u>.
The <u>sun</u> is what <u>has</u> made it <u>so</u>.
When the <u>sun</u> goes <u>down</u> it <u>gives</u> no more <u>warmth</u>
But the <u>ground</u> keeps its <u>warmth</u> down <u>below</u>.

<u>All</u> of the <u>air</u> has some <u>water</u> in<u>side</u> it.
<u>Warm</u> air holds <u>much</u> more than <u>cold</u>.
So <u>when</u> the night <u>air</u> cools <u>down</u> after <u>sun</u>down
That <u>moisture</u> the <u>air</u> cannot <u>hold</u>.

The <u>moisture</u> the <u>air</u> can't hold <u>clings</u> to the <u>grass</u>
Leaving <u>droplets</u> that <u>we</u> call the <u>dew</u>.
But the <u>very</u> next <u>day</u> when the <u>sun</u> come its <u>way</u>
It will <u>dry</u> up the <u>grass</u> and the <u>moisture</u> un<u>do</u>.

CHAPTER THREE
WHAT IS FROST?

All through the day the ground collects warmth
Because of the warmth of the sun.
But when the sun sets on a wintery day,
Freezing, the air then becomes.

The water inside the air starts then to freeze when
The frigid air hits the warm ground.
Frost on a window has a feathery pattern
As crystals are scattered around.

On a very cold night the water condenses
Forming small crystals of ice.
The crystals are white and sparkle like diamonds
That glitter and look really nice.

But to <u>see</u> that you <u>probably</u> will <u>have</u> to rise <u>early</u>
For it <u>melts</u> when the <u>sun</u>shine comes <u>out</u>.
<u>But</u> if you <u>miss</u> it look <u>inside</u> your <u>freezer</u>.
Some <u>frost</u> might be <u>lying</u> a<u>bout</u>.

CHAPTER FOUR
WHY DOES IT SNOW?

When <u>it's</u> very <u>cold</u> down <u>here</u> on the <u>earth</u>
It is <u>freezing</u> way <u>up</u> in the <u>sky</u>.
The <u>water</u> that's <u>trapped</u> in the <u>air</u> turns to <u>ice</u>
And it <u>saves</u> up <u>quite</u> a sup<u>ply</u>.

The <u>water</u> then <u>turns</u> into <u>pretty</u> small <u>crystals</u>
That <u>grow</u> big en<u>ough</u> to then <u>fall</u>.
<u>Soon</u> they're so <u>heavy</u> that <u>even</u> the <u>clouds</u>
Can't <u>hold</u> up the <u>crystals</u> at <u>all</u>.

If the earth's <u>warm</u> when they <u>travel</u> to <u>earth</u>
The <u>snow</u> crystals <u>will</u> simply <u>melt</u>.
<u>But</u> if it's <u>cold</u> down here <u>all</u> of those <u>snowflakes</u>
<u>onto</u> your <u>head</u> will then <u>pelt</u>.

<u>And</u> if down <u>here</u> it stays <u>frigid-y</u> <u>cold</u>
The <u>snow</u> for a <u>while</u> will <u>then</u> stick <u>around</u>
To <u>make</u> it look <u>just</u> like a <u>wintery</u> <u>wonderland</u>
<u>Without</u> the <u>faintest</u> of <u>sounds</u>.

CHAPTER FIVE
WHAT MAKES CLOUDS?

The <u>beautiful</u> clouds are just <u>droplets</u> of water
<u>Coverin</u>g up <u>small</u> bits of <u>dust</u>.
But <u>how</u> did the <u>droplets</u> get <u>up</u> in the <u>sky</u>?
<u>I</u> must know. <u>Yes</u>, I just <u>must</u>!

It's a <u>very</u> long <u>story</u> but I'll <u>tell</u> it to <u>you</u>.
<u>It</u> starts be<u>cause</u> the sun's <u>rays</u> are so <u>hot</u>.
The <u>rays</u> heat the <u>water</u> that's <u>down</u> on the <u>earth</u>.
From <u>puddles</u> to <u>oceans</u> it <u>heats</u> them a <u>lot</u>.

The <u>heat</u> changes <u>water</u> in<u>to</u> water <u>vapor</u>.
The <u>vapor</u> then <u>mixes</u> with <u>air</u>.
The <u>cold</u> air <u>above</u> drops <u>down</u> to the <u>earth</u>
<u>Pushing</u> the <u>warm</u> air up <u>there</u>.

<u>Higher</u> and <u>higher</u> the <u>vapor</u> then <u>rises</u>
<u>Making</u> huge <u>clouds,</u> natura<u>lly</u>.
The <u>vapor</u> then <u>cools</u> and <u>clings</u> to the <u>dust</u>
And <u>gets</u> way too <u>heavy,</u> you <u>see</u>.

It <u>falls</u> to the <u>earth</u> as <u>rain</u>, sleet, or <u>snow</u>
<u>Filling</u> the <u>oceans</u> and <u>then,</u>
The <u>sun</u> comes back <u>out</u> and <u>changes</u> the <u>water</u>
<u>Back</u> into <u>vapor</u> again.

CHAPTER SIX
WHAT MAKES RAIN?

Those <u>gray</u> and black <u>clouds</u> that you
<u>see</u> in the <u>sky</u>
Are <u>clouds</u> that are <u>filled</u> with cold <u>water</u>.
The <u>air</u> all <u>around</u> them is <u>lighter</u> and <u>warmer</u>
And <u>where</u> the drops <u>fall</u> it is <u>hotter</u>.

<u>Downward</u> and <u>downward</u> the <u>rain</u> droplets <u>come</u>
Where they <u>meet</u> and mix <u>up</u> with each <u>other</u>.
They get <u>bigger</u> and <u>bigger</u> as <u>down</u>ward they <u>go</u>
For <u>they</u> join up <u>with</u> one <u>another</u>.

Soon all those raindrops start hitting the earth,
Filling the rivers and oceans.
They pour on our fields, our houses, and streets
In one continuous motion.

One day the droplets will turn into clouds
For the water will change into vapor.
But then the vapor turns back to a cloud
And becomes another rain maker.

CHAPTER SEVEN
WHAT MAKES THUNDER AND LIGHTNING?

In <u>winter</u>time <u>if</u> you should <u>shuffle</u> your <u>feet</u>
On a <u>rug</u> and then <u>go</u> touch some<u>one</u>
There will <u>be</u> a small <u>spark</u> that <u>causes</u> a <u>shock</u>.
You'll <u>laugh</u> and <u>think</u> it great <u>fun</u>.

When <u>you</u> rubbed your <u>feet</u> you were
<u>gaining</u> <u>electrons</u>
As <u>onto</u> you <u>they</u> jump up <u>on</u>.
If the <u>person</u> you <u>touch</u> has
less <u>of</u> them than <u>you</u> have
<u>They</u> will jump <u>over</u> to <u>him</u> and be <u>gone</u>.

At <u>times</u>
you may
<u>see</u> a
<u>faint</u>, little
<u>spark</u>
As it
<u>passes</u>
from <u>you</u>
to your
friend.
But <u>if</u> he
starts
<u>rubbing</u>
his <u>feet</u>
back and

<u>forth</u>
You <u>may</u> get them <u>back</u> once <u>again</u>.

The <u>lightning</u> is <u>like</u> a big <u>spark</u> in the <u>sky</u>
That's <u>filled</u> up with <u>extra</u> el<u>ectr</u>ons.
The earth <u>has</u> less e<u>lectr</u>ons and <u>so</u> by and <u>by</u>
The <u>lightning</u> de<u>posits</u> them <u>and</u> then is <u>gone</u>.

<u>But</u> why should <u>lightning</u> make
<u>such</u> noisy <u>sounds</u>
As <u>electrons</u> go <u>from</u> sky to <u>ground</u>?
Well, the <u>lightning</u> is <u>hot</u>…..I <u>mean</u> **REALLY** <u>hot</u>
And it <u>makes</u> the air <u>hot</u> all <u>around</u>.

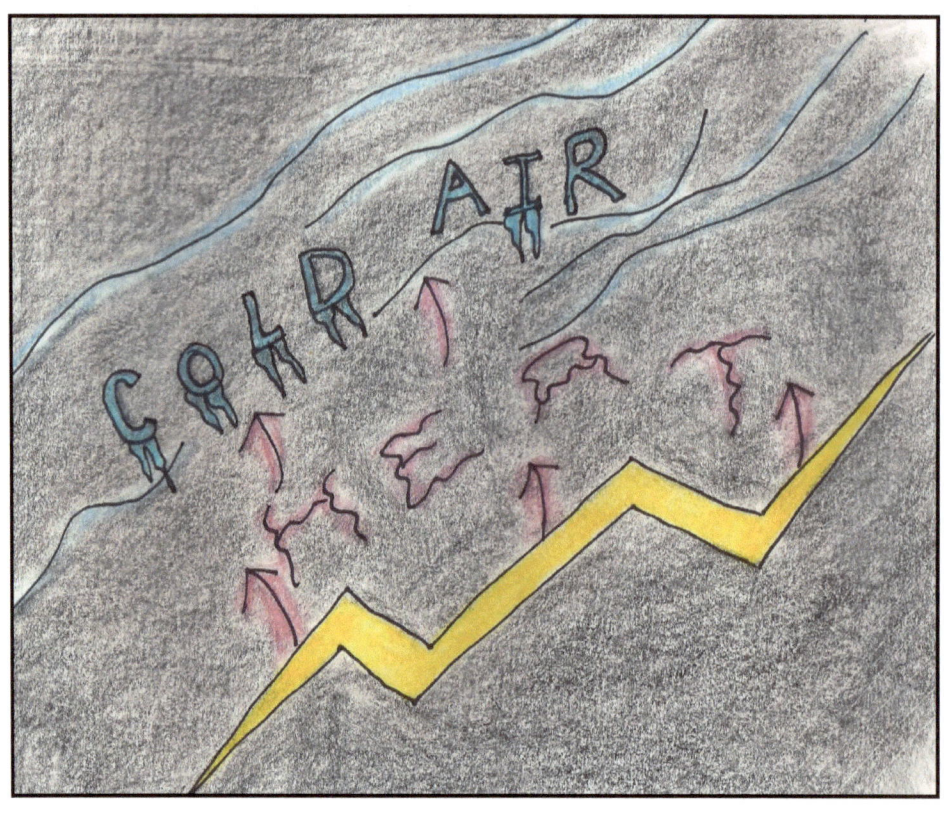

The <u>hot</u> air then <u>spreads</u>, or <u>expands</u> really <u>fast</u>
Bumping <u>into</u> some <u>really</u> cold <u>air</u>
This <u>sets</u> up a <u>great</u> wave of <u>air</u> like a <u>blast</u>
That <u>rolls</u> all <u>around</u> way up <u>there</u>.

The <u>noise</u> it <u>makes</u> is <u>some</u>times quite <u>loud.</u>
It <u>roars</u> and it <u>groans</u> and it <u>thunders.</u>
<u>Lightning</u> bolts <u>flash</u>.
Great <u>waves</u> of air <u>roll.</u>
That <u>we</u> may flinch <u>is</u> little <u>wonder.</u>

CHAPTER EIGHT

WHAT DOES A THUNDERCLOUD LOOK LIKE?

When the <u>sun</u> is <u>shining</u> and there
<u>is</u> a gentle <u>breeze</u>
It <u>seems</u> the weather's <u>smiling</u> and is
<u>wanting</u> kids to <u>please</u>.
But <u>just</u> like people <u>who</u> may have to
<u>frown</u> once in a <u>while</u>
The <u>weather</u> cannot <u>always</u> give small
<u>children</u> a big <u>smile</u>.

16.

It <u>usually</u> rains in <u>thunderstorms</u> but
<u>sometimes</u> it may <u>hail</u>.
The <u>thunder</u> roars and <u>rumbles</u>, and the
<u>wind</u> becomes a <u>gale</u>.
The <u>bottom</u> of a <u>thunder</u>storm is
<u>flat</u> and gray and <u>dark</u>.
At the <u>top</u> it is quite <u>puffy</u> and it
<u>forms</u> a kind of <u>arch</u>.
<u>It's</u> great fun to <u>watch</u> a thunder<u>storm</u> if
you're in<u>side</u>
For you'll <u>get</u> to see a <u>great</u> show like ones
<u>on</u> Fourth of <u>July</u>.

CHAPTER NINE
WHAT ARE HAILSTONES?

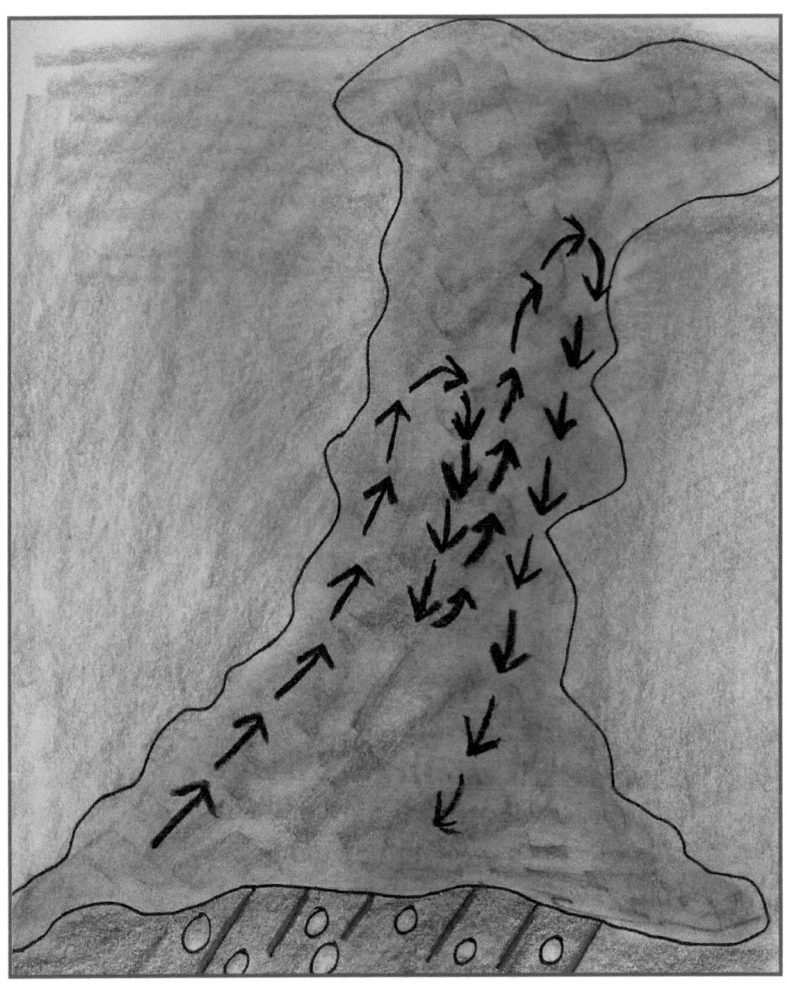

Hail starts out like rain although the
rain just may not hold.
As it falls it's tossed back up
Into clouds that are cold.

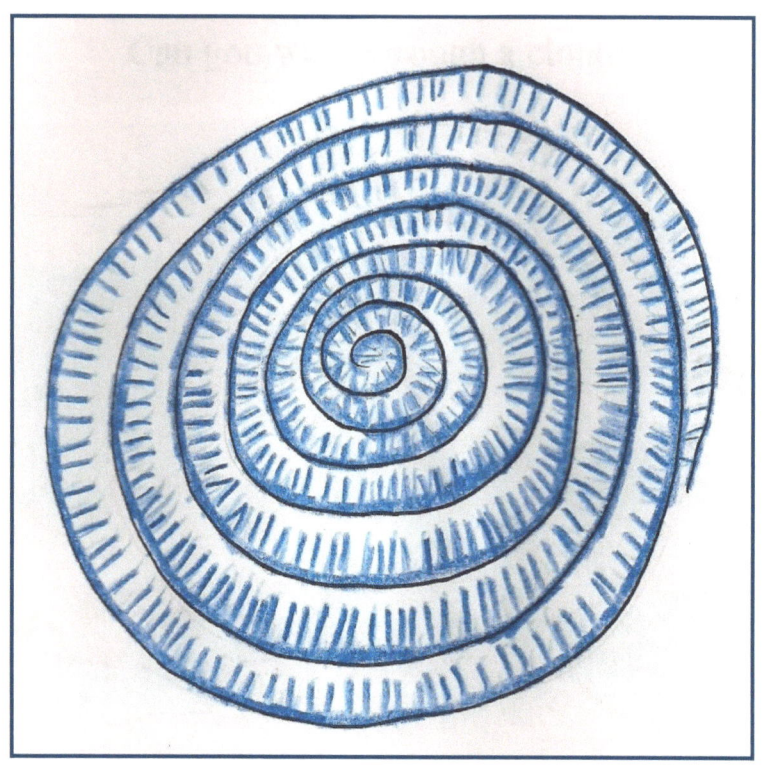

Up and down like an elevator
it rides on the breeze.
When it's tossed up high it's cold.
It soon begins to freeze.
When it comes down it starts to melt but
up again it goes,
Up and down and up and down
as the icy crystal grows
But when the crystal gets too big it
can no longer stay
So crashes down onto the ground
because too much it weighs.

CHAPTER TEN
CAN YOU WALK THROUGH A CLOUD?

If you've <u>ever</u> been on an <u>airplane</u> with white
<u>puffy</u> clouds be<u>low</u>
That <u>kind</u> of looked like <u>cotton</u> candy
or <u>mounds</u> of fluffy <u>snow</u>
Then you <u>may</u> have wished to <u>jump</u> and play
on <u>clouds </u>within your <u>view</u>.
But, of <u>course</u>, a cloud can't <u>hold</u> you up.
You <u>simply</u> would fall <u>through</u>.

20.

But there <u>are</u> some clouds you <u>can</u> walk through
although not in the <u>sky</u>.
It's the <u>fog</u> you find in the <u>early</u> morning.
Just <u>pretend</u> you're walking <u>high</u>.
You can <u>also</u> breathe <u>into</u> cold air and
<u>form</u> a cloud that <u>stays,</u>
<u>Only</u> for a <u>moment</u> for it <u>quickly</u> fades <u>away</u>.

CHAPTER ELEVEN
WHAT MAKES THE WIND BLOW?

Cold <u>air</u> is thick and <u>heavy</u>.
Hot <u>air</u> is thin and <u>light</u>
When <u>cold</u> air meets hot <u>air</u> the hot air's
<u>shot</u> up out of <u>sight</u>.
The <u>cold</u> air is now <u>down</u> by us.
The <u>warm</u> air's in the <u>sky</u>.
As the <u>warm</u> air cools it <u>drops</u> back down.
Cold <u>air</u> that warms will <u>rise</u>.

So, the <u>air</u> is constantly <u>moving</u>
Starting <u>off</u> as a gentle <u>breeze</u>.
But at <u>times</u> the winds will <u>whip</u> and blow
All <u>around</u>, and it makes us <u>freeze</u>.

CHAPTER TWELVE
WHEN DOES AIR MOVE THE FASTEST?

Dark things soak in heat from the sun.
White things turn rays away.
Things like the asphalt on driveways and roads
Heat the air faster than things in the shade.

That's why you get hot when playing outdoors
In the summer and head for the lake.
The rays from the sun heat the water up slowly.
But the land gets so hot you can
about fry an egg!

When the <u>land</u> gets <u>hot</u>, it <u>heats</u> the air <u>above</u> it.
That <u>air</u> rises <u>at</u> a speeding <u>pace</u>.
The <u>cooled</u>-off air that's <u>in</u> the shade
<u>or</u> that's on <u>water</u>
<u>Rushes</u> in to <u>take</u> up all that s<u>pace</u>.

The <u>greater</u> that the <u>difference</u> is
bet<u>ween</u> the hot and <u>cold,</u>
The <u>faster</u> that the <u>air</u> can then re<u>treat</u>.
The <u>faster</u> that the <u>cold</u> air rushes
<u>to</u> the air that's <u>hot</u>
The <u>windier</u> the <u>day</u> that you will <u>greet</u>.

CHAPTER THIRTEEN
WHERE DO HURRICANES COME FROM?

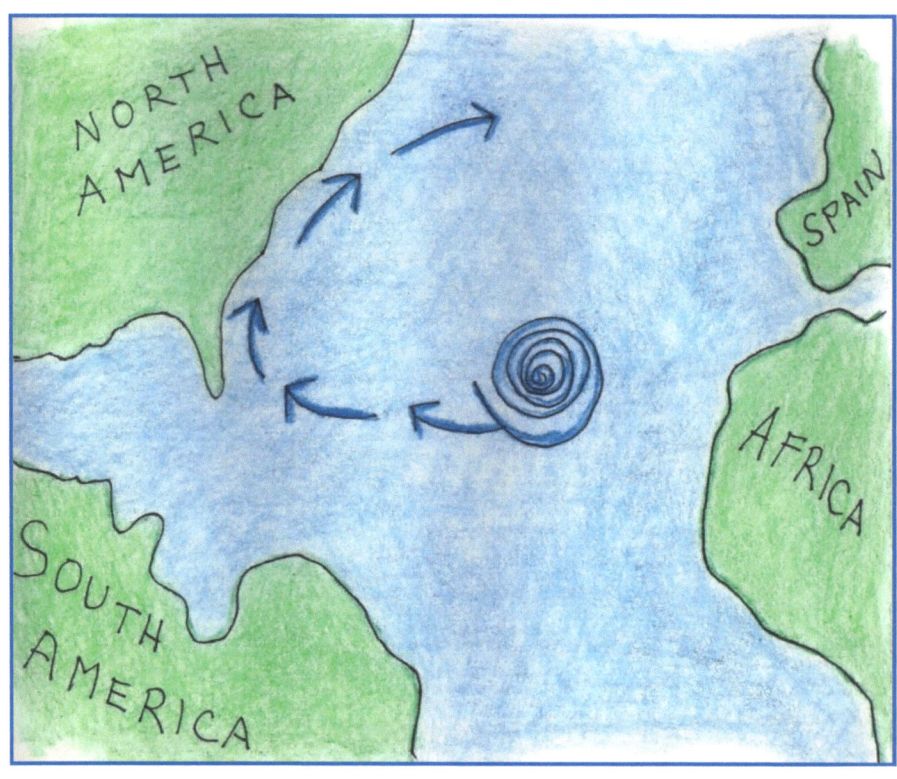

Hurricanes are <u>born</u> over <u>warm</u>, tropical <u>water</u>.
The <u>sun</u> beats down on <u>water</u> all the <u>day</u>
<u>Warm</u>ing up the <u>air</u> and <u>picking</u> up the <u>moisture</u>
<u>As</u> the air <u>travels</u> along its <u>way</u>.

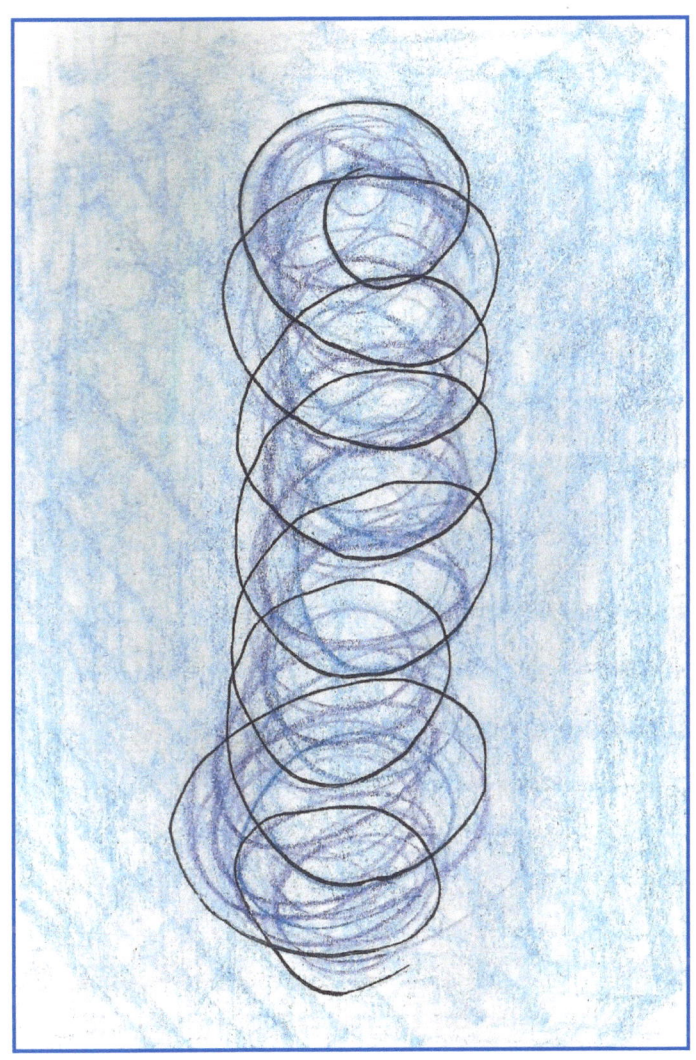

As the air moves upward, the cold air all around it
Quickly moves in and takes up the space.
The cold air then will push the hot air
up a little higher,
And man! It's blowing at an incredible pace!

<u>Around</u> and <u>around</u> that
<u>fierce</u> wind keeps on <u>whirling</u>
<u>Except</u> right in the <u>center</u>.
They <u>call</u> this space the <u>eye</u>.
The <u>rain</u> seems to <u>beat</u> down in a <u>horrible</u> <u>hurry</u>
But <u>there</u> in the <u>center</u> you can <u>see</u> blue <u>sky</u>!

When the <u>eye</u> passes <u>over</u> you
you <u>know</u> you'd best get <u>ready</u>.
For the <u>hurricane's</u> other <u>half</u> will come your <u>way</u>.
But <u>weathermen</u> can <u>tell</u> just where a
hurricane is <u>going</u>
And <u>you</u> can pack your <u>bags</u> because you
<u>do</u> not have to <u>stay</u>.

CHAPTER FOURTEEN
HOW DO CLOTHES KEEP YOU WARM?

Children wear warm coats or jackets, mittens,
gloves and hats
When the air gets cold in wintertime.
And I guess you figure that these
things will keep cold out.
That's why to bundle up you are inclined.

But <u>actua</u>lly the <u>opp</u>osite is <u>what</u> is really <u>true</u>.
It's <u>because</u> the leather, <u>wool</u> or fur you <u>wear</u>
Is <u>full</u> of deepened <u>spaces</u> that will
<u>trap</u> your body <u>heat,</u>
Not <u>let</u>ting it e<u>scape</u> into the <u>air</u>.

So, your <u>warm</u> clothes aren't <u>there</u> so that it <u>keeps</u>
the cold air <u>out</u>
<u>But</u> it's there to <u>keep</u> all your heat <u>in</u>.
And in <u>summer</u> when you <u>want</u> to let the
<u>body</u> heat <u>out</u>
<u>You</u> wear clothes that <u>are</u> skimpy and <u>thin.</u>

CHAPTER FIFTEEN
WHY DO YOU WEAR LIGHT COLORED CLOTHING IN THE SUMMER?

In the <u>winter</u> you wear <u>lots</u> of clothes that
<u>usually</u> are <u>heavy</u>
And <u>not</u> clothes that are <u>flimsy</u> ones or <u>thin</u>.
Most <u>of</u> the time you <u>will</u> find that the
<u>colors</u> are quite <u>dark</u>.
It's be<u>cause</u> dark colors <u>soak</u> up heat with<u>in</u>.

32.

But in summer <u>you</u> wear just as <u>little</u> as you <u>can</u>
And <u>usually</u> lighter <u>colors</u> are worth<u>while.</u>
Not <u>to</u> wear black or <u>navy</u> blue would
<u>surely</u> be your <u>plan</u>
But <u>not</u> so you can <u>wear</u> the latest <u>style.</u>

Of <u>course</u> there is a <u>reason</u>, and it
<u>is</u> quite a good <u>one</u>.
You <u>don't</u> wear clothes that <u>gather</u> all the <u>heat.</u>
Light <u>colors</u> will re<u>flect</u> most of the <u>rays</u> of the <u>sun</u>
So you <u>will</u> not get so <u>hot</u> when at the <u>beach.</u>

CHAPTER SIXTEEN
WHY DO WINDOWS STEAM UP?

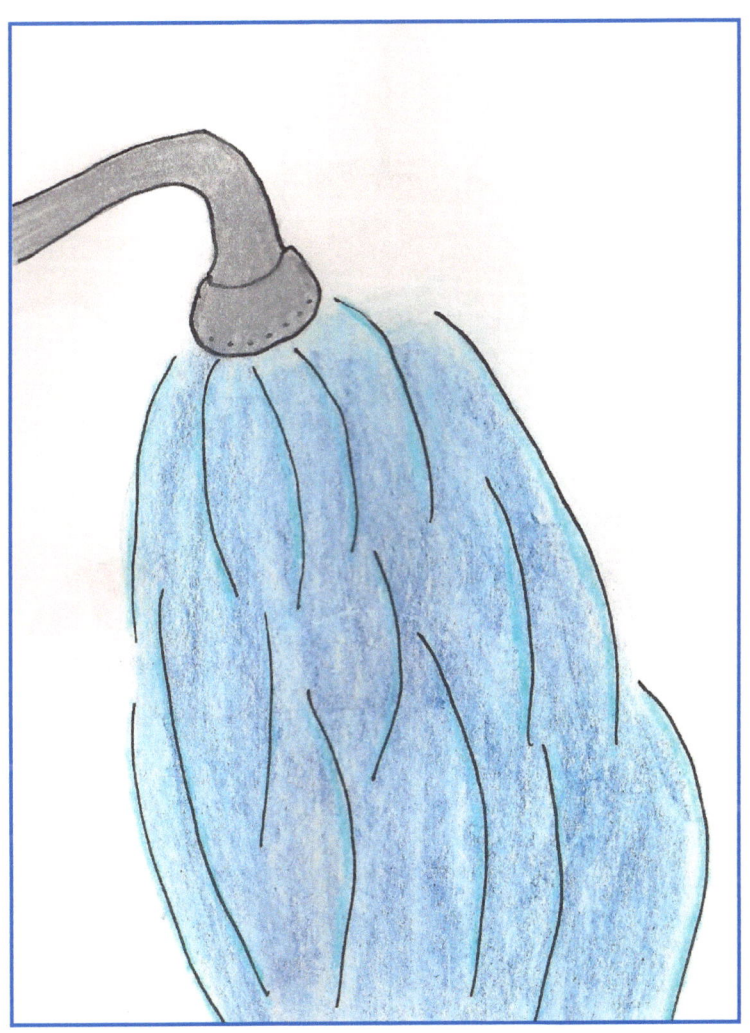

When you <u>hop</u> into the <u>shower</u>
the <u>room</u> gets hot and <u>steamy</u>.
The <u>mirrors</u> get all <u>foggy</u> and it
<u>looks</u> so nice and <u>dreamy</u>.

When <u>some</u>thing very <u>good</u> is boiling <u>in</u> your
mother's <u>pot</u>
The <u>windows</u> will cloud <u>up</u>.
You <u>barely</u> can see <u>out</u>.

Or when <u>riding</u> in your <u>car,</u> when you
<u>had</u> not much to <u>do,</u>
Did you <u>breathe</u> hard on the <u>window</u> and then
<u>pretty</u> pictures <u>drew?</u>

Well, the <u>air</u> that's near those <u>windows</u>
was <u>very</u> wet and <u>hot</u>.
When <u>hot</u> air hits cold <u>windows</u>
the <u>air's</u> cooled off a <u>lot</u>.

Cooled <u>air</u> can't hold the <u>moisture</u>.
It con<u>den</u>ses on the <u>spot</u>.
So the <u>mirrors</u> and <u>windows</u>
Get <u>clouded</u> up a lot.

CHAPTER SEVENTEEN
WHY DO YOU SEE YOUR BREATH ON A COLD DAY?

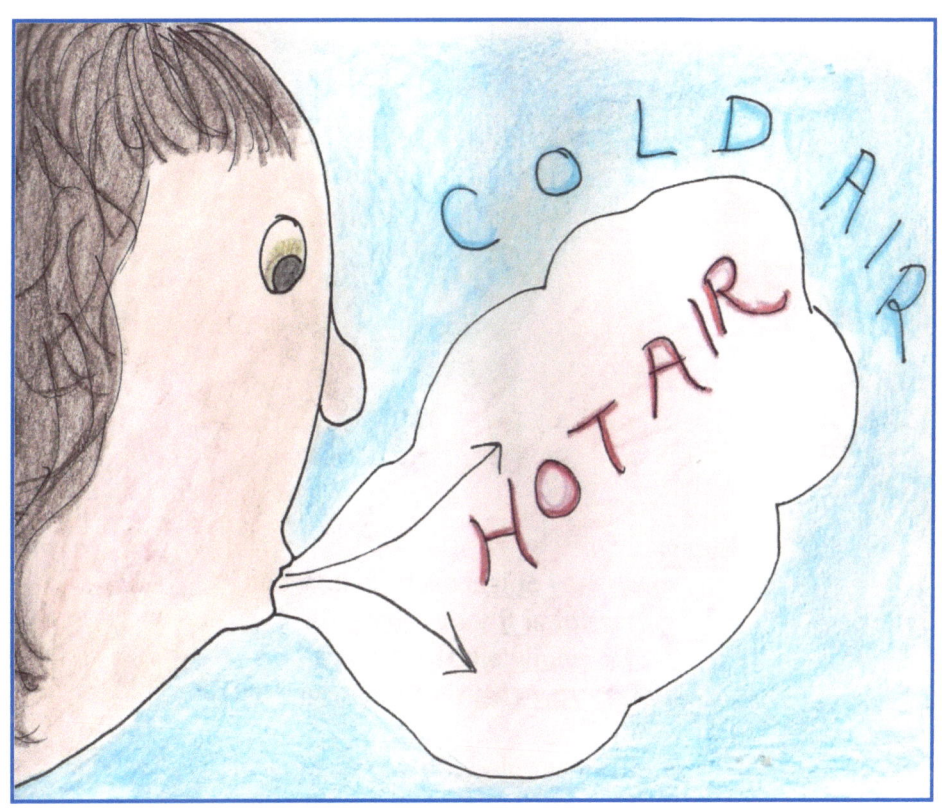

When the air is cold and frosty
Your breath's like smoke floating by.
When it leaves your mouth it's misty.
Can you tell the reason why?

38.

The <u>warm</u> air deep in<u>side</u> your lungs,
With <u>water</u> vapor <u>they</u> are filled.
And <u>when</u> it hits the <u>winter</u> cold
That <u>warm</u> moist air gets <u>very</u> chilled.

The <u>cold</u> air cannot <u>hold</u> the moisture
<u>So</u> the moisture s<u>queezes</u> out.
We <u>say</u> the moisture <u>is</u> condensing.
<u>And</u> it's true so <u>have</u> no doubt.

But <u>I</u> think that it's <u>much</u> more fun
A <u>dragon</u> to pre<u>tend</u> to be.
So <u>you</u> can breathe out <u>smoke</u> and fire
For <u>all</u> your little <u>friends</u> to see.

CHECK OUT OTHER BOOKS FROM THE
SCIENCE WITH A BEAT
SERIES

BOOK ONE: OUT BEYOND

BOOK TWO: OUR WONDERFUL PLANET
EARTH

BOOK THREE: WEATHER

BOOK FOUR: WATER WORLD

BOOK FIVE: WEATHER